This template is designed to help a farm prepare for an audit under the Harmonized Standard, conducted by Equicert or another third party auditor. The contents are copyright 2013, by Equicert, except material from the standard itself, which is the intellectual property of the United Fresh Produce Association. This standard is poised to become the new industry standard for the future of fresh produce growing in North America. Following the requirements of this standard should put any farm very close to the requirements of any other standard, as most farm food safety standards have 90% of their content in common and the Harmonized Standard was designed to feature the best and strongest features of existing popular standards to avoid multiple audits.

It must be cautioned that following this template will not make your produce safer, neither is produce that is produced in conformity with any standard necessarily safer than other produce. No food safety standard can guarantee that fresh produce is not "contaminated" with bacteria or other pathogens. Produce is grown on farms, of course, which are environments that are conducive for the growth of living things such as bacteria. There is no such thing as a sterile farm or packinghouse, and there is no such thing as risk free fresh produce. The best we can do is to reduce avoidable risks by using the safest practices known.

A farm food safety program is composed of four parts. Your food safety policy, a written statement of your view of food safety on the farm. Second, your food safety plan, which includes risk assessments of the various physical, chemical, and biological risks on your farm. And SOP's, Standard Operating Procedures, which are work instructions that describe how the work on the farm gets done. Finally, there is the implementation of that program, which is simply how you put that plan into action. I encourage you to fill out this template honestly, so that it accurately reflects the practices on your farm.

If you are interested in food safety materials for group audits, we also have a food safety group template available through your Equicert auditor, or it can be ordered through your local bookstore. A quick reference copy of the 2013 Harmonized Standard is provided at the end, courtesy of the United Fresh Produce Association. A request for audit is provided on the last page.

Michael Hari
Equicert
5467 S 800 W
Delphi, IN 46923-8823
(765) 589 3058

Good Agricultural Practices Food Safety Plan

for

_____
Name of Farm

Food Safety Policy

_____

_____

_____

_____

_____

_____

_____

_____

Signature

_____

Date

Farm Information

Farm name and mailing address:

Site address:

Person responsible for food safety program at this location:

Name/ Position

Address

Telephone:

Fax:

E-mail

Field Map Or Description

Here attach a farm map, aerial photo, or a written description of the farm or packing house that includes toilets, sinks, handwashing stations, fields (with designation if appropriate), portable toilet locations, septic systems and sewer lines, water lines, manure storage and livestock, packing lines, inside water lines, risks from adjacent properties, and uses of land and buildings as applicable. Hand sketches are fine.

_____

_____

_____

_____

_____

_____

_____

Use this blank page, and all additional blank pages in the manual for sketches or record additional information.

Water Management Plan

Here write your procedures for testing, risk assessment, sampling of water (if you use a lab for testing, do they give sample instructions? If so attach.) , and a description of your water system. Include such details as your maintenance schedule, water bills, municipal water tests, irrigation water tests, pond/river risk assessment, as applicable. What do you do if you fail a water test?

_____

_____

_____

_____

_____

_____

_____

Water Management Plan Contd.

Manure and Compost/Animal Control

Here explain how any manure or compost is treated (eg time, temperature, and other treatment to reduce pathogen risk). Do you wait before applying raw manure? If so explain here. Explain also any procedures used to reduce risk from your own or neighboring domestic animals, wild animals, or feral animals. Explain your procedure for monitoring for wild/feral animals. Do you use work animals in the field? If so what do you do to reduce risk from domestic draft animals. Include details like field lanes, buffer strips, manure handling, etc.

_____

_____

_____

_____

_____

_____

_____

Worker Health, Hygiene, and Sanitary Practices

EQUICERT

Here explain what policies and procedures, if any, you have for toilet, handwashing, toilet paper disposal, and sick employees handling produce.

_____

_____

_____

_____

_____

_____

_____

_____

_____

Harvesting, Field Packing, and Transportation

Describe here how you harvest, pack and transport your produce. Include a list of all equipment that contacts produce. Give a detailed SOP for each commodity that includes the method of handling, harvesting, and packing including your choice of packing materials. Do you have documentation for your packing material suitability for food contact? Please attach it.

_____

_____

_____

_____

_____

_____

_____

_____

Harvest, Field Pack, Packing and Transportation Contd.

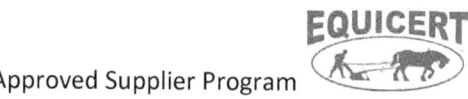

Approved Supplier Program

Describe here how you select sources for packaging, inputs like chemicals and fertilizers, seeds, etc. List your suppliers with contact info.

_____

_____

_____

_____

_____

_____

_____

_____

_____

_____

_____

Training

Here explain how often and how employees are trained. What language are employees trained in?

___

___

___

___

___

___

___

___

___

___

EQUICERT

Traceability and Recall

Please explain how your traceback system works. Can you trace produce one step forward? Do you ever mix produce lots? Can you identify in writing the source of everything that comes in contact with produce such as chemicals, fertilizers, compost, manure, post harvest chemical treatments, packaging, harvest containers, tools, and equipement? Do you do a mock recall? Where is the record of your mock recall? How soon can you find produce that has left your farm/facility?

_____

_____

_____

_____

_____

_____

_____

Produce Contact Equipment List

Here list everything that comes in contact with produce.

_____

_____

_____

_____

_____

_____

_____

_____

_____

_____

Training Log

List here the name, date, and type of training for each employee.

Addendum #1: United Fresh Produce Association Harmonized Standard

Pre Farm Gate Standard: (Reproduced by permission—for brevity's sake only the requirement and procedure are reproduced here.)

	Requirement	Procedure
1.		
1.1.		
1.1.1.	A food safety policy shall be in place.	A written policy shall outline a commitment to food safety, in general terms, how it is implemented and how it is communicated to employees, and be signed by Senior Management.
1.1.2.	Management has designated individual(s) with roles, responsibilities and resources for food safety functions.	The food safety plan shall designate who has the responsibility and authority for food safety, including a provision for the absence of key personnel. Twenty-four hour contact information shall be available for these individuals in case of food safety emergencies. The organization's senior management shall determine and provide, in a timely manner, the resources needed to implement and maintain the food safety plan.
1.1.3.	There is a disciplinary policy for food safety violations.	There shall be a policy that establishes corrective actions for personnel who violate established food safety policies or procedures.
1.2.		
1.2.1.	There shall be a written food safety plan that covers the operation.	The food safety plan shall identify all locations of the operation and products covered by the plan. The plan shall address potential physical, chemical, and biological hazards and hazard control procedures, including monitoring, verification and recordkeeping, for the following areas: water, soil amendments, field sanitation, production environment, and worker practices.
1.2.2.	The food safety plan shall be reviewed at least annually.	Operation shall be responsible for reviewing their food safety plan at least annually, documenting the review procedure and revising the plan as necessary. Updated or revised on date shall be indicated.
1.3.		
1.3.1.	Documentation shall be kept that demonstrates the food safety plan is being followed.	Documents and records of procedures, standard operation procedures (SOPs) and policies shall be in place for meeting each of the food safety standards identified in the Food Safety Plan.

1.3.2.	Documentation shall be readily available for inspection.	Documents and records may be maintained on-site or at an off-site location, or accessible electronically (e.g., MSDS), and shall be available for inspection in a reasonable timeframe or as required by prevailing regulation.	
1.3.3.	Documentation shall be retained for a minimum period of two years, or as required by prevailing regulation.	Document and record handling policy or procedures require that documentation required by the food safety plan shall be retained for a minimum of two years, or as required by prevailing regulation.	
1.4.			
1.4.1.	All personnel shall receive food safety training.	All personnel shall receive training in the food safety policy and plan, food safety procedures, sanitation and personal hygiene appropriate to their job responsibilities. Personnel shall receive training at hire and refresher training at prescribed frequencies. Documentation of training is available.	
1.4.2.	Personnel with food safety responsibilities shall receive training sufficient to their responsibilities.	The individual designated for food safety responsibilities demonstrates knowledge of food safety principles. Food safety designate has completed at least one formal food safety course/workshop or by job experience.	
1.4.3	Subcontractors are held to the relevant food safety standards as they would be as employees.	Operations shall have procedures and/or records to demonstrate that subcontractors whose activities can affect food safety have been informed of and, to the extent that can be verified, are in compliance with the relevant requirements of the Field Operations and Harvesting standards.	
1.5.			
1.5.1.	Where laboratory analysis is required in the Food Safety Plan, testing shall be performed by a GLP laboratory using validated methods.	Operation utilizes laboratories that have, at minimum, passed a Good Laboratory Practices (GLP) audit or participates in a Proficiency Testing program, and utilizes BAM, AOAC International or testing methods that have been validated for detecting or quantifying the target organism(s) or chemical(s).	

1.5.2.	Where microbiological analysis is required in the food safety plan, samples shall be collected in accordance with an established sampling procedure.	Operation utilizes a written sampling protocol when collecting samples for microbiological testing.
1.5.3.	Tests, their results and actions taken must be documented.	All results for microbiological testing required in the Operation's food safety plan shall be recorded and the records maintained for two years.
1.5.4.	All required testing shall include test procedures and actions to be taken based on the results.	For all microbiological testing required by the food safety plan, Operation has a written testing procedure that includes test frequency, sampling, test procedures, responsibilities and actions to be taken based on results. If finished product is tested for pathogens or other adulterants, Operation's procedures require that it shall not be distributed outside the operation's control until test results are obtained.
1.6.		
1.6.1.	A documented traceability program shall be established.	Records that enable reconciliation of product delivered to recipients (one step forward) shall be maintained except for direct to consumer sales. Records shall be maintained that link product with source of the produce or production inputs, e.g., soil amendments, fertilizers, seeds/transplants, agricultural chemicals, homemade preparations (one step backward). Records shall include the date of harvest, quantities, farm identification (field or block), transporter and non-transporter. Additional information may be included. Contents and retention of records shall be consistent with applicable regulations.
1.6.2.	A trace back and trace forward exercise shall be performed at least annually.	The trace back and trace forward exercise shall achieve accurate traceability within 4 hr or as required by applicable regulations. Trace exercise shall achieve 100% reconciliation of product to recipients.
1.7.		
1.7.1.	A documented recall program, including written procedures, shall be established.	The recall program shall have a designated recall team. A mock recall exercise shall be performed at least annually at the operation being audited. The mock recall shall include the trace back and trace forward exercise and shall be completed as stated in the program and in compliance to applicable regulations.
1.8.		

1.8.1.		The operation shall have documented corrective action procedures.	A documented Corrective Action is required for an observation or audit that contains a non-conformance with food safety requirements. The responsibility, methods, and timelines to address Corrective Actions shall be documented and implemented.
1.9.			
	1.9.1.	The operation shall have documented self-audit procedures.	Internal audits will be conducted at a minimum annually by an assigned individual utilizing this standard to assist in the self-audit. All aspects of the Operation's food safety plan will be audited and a written record of required corrective action will be documented.
2.			
2.1.			
	2.1.1.	The food safety plan shall, initially and at least annually thereafter, evaluate and document the risks associated with land use history and adjacent land use including equipment and structures.	When land use history or adjacent land use indicates a possibility of physical, chemical or biological contamination, preventive controls shall be performed and documented to mitigate food safety risk. The assessment is re-performed, and documented, at least annually for environmental conditions or risk awareness that has changed since the last assessment. The assessment shall include indoor growing facilities and structures such as green houses and hydroponics.
	2.1.2.	For indoor growing and field storage facilities, facility shall be designed, constructed and maintained in a manner that prevents contamination of produce.	Facility and equipment structures and surfaces (floors, walls, ceilings, doors, frames, hatches, etc.) shall be constructed in a manner that facilitates cleaning and sanitation and does not serve as harborage for contaminants or pests. Chill and cold storage loading dock areas shall be appropriately sealed, drained and graded. Fixtures, ducts, pipes and overhead structures shall be installed and maintained so that drips and condensation do not contaminate produce, raw materials or food contact surfaces. Water from refrigeration drip pans shall be drained and disposed of away from product and product contact surfaces. Drip pans and drains shall be designed to assure condensate does not become a source of contamination. Air intakes shall not be located near potential sources of contamination.
2.2.			
	2.2.1.	Operation shall have a policy for toilet, hygiene, and health.	Each operation shall establish written policies for their specific operations, which shall be in compliance with prevailing regulations for Worker Health and Hygiene Practices.

2.2.2.	Employees and visitors shall follow all personal hygiene practices as designated by the operation.	Operation's hygiene policies shall apply to all employees, contractors, visitors, buyers, product inspectors, auditors, and other personnel in the field. The operation shall designate competent supervisory personnel to ensure compliance by all workers, visitors, and field personnel with the requirements in this section.
2.2.3.	Toilet facilities shall be designed, constructed, and located in a manner that minimizes the potential risk for product contamination and are directly accessible for servicing.	Toilet and handwashing facilities are situated during operation and servicing, and maintained so as not to pose a hazard to the produce or other opportunity for contamination.
2.2.4.	Toilet facilities shall be of adequate number, easily accessible to employees and in compliance with applicable regulation.	The operation will have verification that the number of toilet facilities and their location relative to employees meets the more stringent of federal, state or local regulations.
2.2.5.	Toilet and wash stations shall be maintained in a clean and sanitary condition.	Toilet paper shall be available in toilet facility. Wash stations shall be located with the field sanitation units and include hand wash facilities with water that meets the microbial standard for drinking water, hand soap, disposable towels or other hand drying device, towel disposal container, and a tank that captures used hand wash water for disposal. These stations shall be provided inside or adjacent to toilet facilities.
2.2.6.	Personnel shall wash their hands at any time when their hands may be a source of contamination.	Personnel shall wash their hands prior to start of work, after each visit to a toilet, after using a handkerchief/tissue, after handling contaminated material, after smoking, eating or drinking, after breaks and prior to returning to work and at any other time when their hands may have become a source of contamination.
2.2.7.	Signage requiring handwashing is posted.	Signage in applicable languages and/or pictures shall be provided adjacent to hand wash facilities requiring people to wash their hands after each toilet visit.

2.2.8.	Clothing, including footwear, shall be effectively maintained and worn so as to protect product from risk of contamination.	Operation shall have a policy that employee clothing shall be clean at the start of the day and appropriate for the operation.
2.2.9.	If gloves are used, the operation shall have a glove use policy.	If rubber, disposable, cloth or other gloves are used in contact with product, the operation shall have a glove use policy that specifies how and when gloves are to be used, cleaned, replaced and stored. Policy shall be in compliance with current industry practices or regulatory requirements for that commodity.
2.2.10.	Protective clothing, when required, shall be maintained, stored, laundered and worn so as to protect product from risk of contamination.	If protective clothing is used in proximity to product, the operation shall have a policy or procedures for how and when protective clothing are to be used, cleaned, replaced and stored. Policy shall be in compliance with current industry practices or regulatory requirements for that commodity.
2.2.11.	When appropriate, racks and/or storage containers or designated storage area for protective clothing and tools used by employees shall be provided.	When employees wear protective clothing, such as aprons and gloves, the Operation shall have a policy for how the clothing and tools shall be stored when not in use so as to avoid potential contamination.
2.2.12.	The wearing of jewelry, body piercings and other loose objects (e.g. false nails) shall be in compliance to company policy and applicable regulation.	Operation shall have a policy that personal effects such as jewelry, watches or other items shall not be worn or brought into fresh fruit and vegetable production areas if they pose a threat to the safety and suitability of the food. Policy shall be in compliance with current industry practices or regulatory requirements for that commodity.
2.2.13.	The use of hair coverings shall be in compliance to company policy and applicable regulation.	The Operation shall have a policy that addresses use of hair coverings (e.g., hair nets, beard nets, caps), which is in compliance with prevailing regulation.
2.2.14.	Employees' personal belongings shall be stored in designated areas.	Operation shall have a policy for when and how employee's personal belongings shall be stored so as not to be a source of product contamination.

2.2.15.	Smoking, chewing, eating, drinking (other than water), urinating, defecating or spitting is not permitted in any growing areas.	Operation shall have policy prohibiting smoking, eating, chewing gum or tobacco, drinking other than water except in designated areas. Such areas shall be designated so as not to provide a source of contamination. Operation shall have policy prohibiting urinating or defecating in any growing area.
2.2.16.	Operation shall have a written policy that break areas are located so as not to be a source of product contamination.	Break areas shall be designated and located away from food contact/handling zones and production equipment.
2.2.17.	Drinking water shall be available to all field employees.	Drinking water, which meets drinking water standards, shall be easily accessible to field personnel and in compliance with applicable regulation. Bottled water or potable drinking water stations with single-use cups and a trash receptacle shall be available to all field employees.
2.2.18.	Workers and field personnel who show signs of illness shall be restricted from direct contact with produce or food-contact surfaces.	Operation shall have a written policy that restricts personnel who show signs of illness (e.g., vomiting, jaundice, diarrhea) from contact with product or food contact surfaces. Policy shall require that any person so affected immediately report illness or symptoms of illness to the management.
2.2.19.	Personnel with exposed cuts, sores or lesions shall not be engaged in handling product.	Minor cuts or abrasions on exposed parts of the body are acceptable if covered with a non-permeable covering, bandage or glove. Bandages on hands shall be covered with gloves in compliance with operation's glove policy.
2.2.20.	Operation shall have a blood and bodily fluids policy.	There shall be a written policy specifying the procedures for the handling/ disposition of food or product contact surfaces that have been in contact with blood or other bodily fluids.
2.2.21.	First aid kits shall be accessible to all personnel.	The kits shall be readily available in the vicinity of field work and maintained in accordance with prevailing regulation. The kit materials shall be kept in a sanitary and usable condition.
2.3.		

2.3.1.	Use of agricultural chemicals shall comply with label directions and prevailing regulation.	Agricultural chemicals, including post-harvest chemicals such as biocides, waxes and plant protection products, must be registered for such use as required by prevailing regulation, and used in accordance with label directions including application rates, worker protection standards, personal protection equipment, container disposal, storage, and all requirements specified for the chemical or compound. Records of agricultural use are maintained.
2.3.2.	If product is intended for export, agricultural chemical use, including post-harvest chemicals, shall consider requirements in the intended country of destination.	The operation shall have procedures, such as pre-harvest interval and application rate, sufficient to meet the MRL entry requirements of the country(ies) in which the product is intended to be traded, if known during production.
2.3.3.	Agricultural chemicals shall be applied by trained, licensed or certified application personnel, as required by prevailing regulation.	Operation maintains records demonstrating that all personnel responsible for chemical applications are trained and/or licensed, or supervised by licensed personnel, in compliance with prevailing regulation.
2.3.4.	Water used with agricultural chemicals shall not be a source of product or field contamination.	Water used to dilute or deliver agricultural chemicals shall be from a source in compliance with the Water System Risk Assessment and Water Management Plan, consistent with current industry practices or regulatory requirements for that commodity.
2.3.5.	Agricultural chemical disposal shall not be a source of product or field contamination.	Operation shall have procedures for disposal of waste agricultural chemicals and for cleaning of application equipment that protects against contamination of product and growing areas.
2.4.		
2.4.1.		

2.4.1.1.	A water system description shall be available for review.	Water sources and the production blocks they may serve shall be documented and current. The description shall include one or more of the following: maps, photographs, drawings (hand drawings are acceptable) or other means to communicate the location of water source(s), permanent fixtures and the flow of the water system (including holding systems, reservoirs or any water captured for re-use). Permanent fixtures include wells, gates, reservoirs, valves, returns and other above ground features that make up a complete irrigation system shall be documented in such a manner as to enable location in the field.
2.4.1.2.	The water source shall be in compliance with prevailing regulations.	Agricultural water shall be sourced from a location and in a manner that is compliant with prevailing regulations.
2.4.1.3.	Agricultural water systems shall not be cross-connected with human or animal waste systems.	Water systems intended to convey untreated human or animal waste shall be separated from conveyances utilized to deliver agricultural water.
2.4.2.		
2.4.2.1.	An initial risk assessment shall be performed and documented that takes into consideration the historical testing results of the water source, the characteristics of the crop, the stage of the crop, and the method of application.	A review or new assessment shall be conducted seasonally and any time there is a change made to the system or a situation occurs that could introduce an opportunity to contaminate the system. The risk assessment shall address potential physical, chemical, and biological hazards and hazard control procedures for the water distribution system.
2.4.3.		
2.4.3.1.	There shall be a water management plan to mitigate risks associated with the water system on an ongoing basis.	The water management plan shall include the following: preventive controls, monitoring and verification procedures, corrective actions, and documentation. The plan shall be reviewed following any changes made to the water system risk assessment and adjusted accordingly to incorporate such changes. Training and/or retraining of personnel having oversight or performance duties shall be documented.

2.4.3.2.	Water testing shall be part of the water management plan, as directed by the water risk assessment and current industry standards or prevailing regulations for the commodities being grown.	As required, there shall be a written procedure for water testing during the production and harvest season, which includes frequency of sampling, who is taking the samples, where sample is taken, how the sample is collected, type of test and acceptance criteria. If all agricultural water is sourced from a municipal source, the municipal testing shall suffice. The frequency of testing and point of water sampling shall be determined based on the risk assessment and current industry standards for commodities being produced.
2.4.3.3.	The testing program shall be implemented consistent with the water management plan.	Testing shall be performed and documented according to procedures described in the water management plan.
2.5.		
2.5.1.	The operation has a written risk assessment on animal activity in and around the production area.	There shall be a written assessment of the growing fields and adjacent land, prior to each growing season, focusing on domestic and wild animal activity including grazing and feeding operations, noting crop characteristics, type and approximate number of animals, proximity to the growing field, water sources, and other relevant factors.
2.5.2.	The operation routinely monitors for animal activity in and around the growing area during the growing season.	There shall be scheduled monitoring of growing fields and adjacent land for evidence of animal activity. A frequency of monitoring and assessment shall be established based on production factors, such as the crop, geography, and other conditions.
2.5.3.	Based on the risk assessment, there shall be measures to prevent or minimize the potential for contamination from animals, including domestic animals used in farming operations.	The operation shall have risk-appropriate actions to prevent or minimize the potential for contamination of produce with pathogens from animal feces, including from domestic animals used in farming operations. There shall be a written record of any mitigation or corrective actions. Preventive measures and corrective actions shall comply with all local, state and federal regulations concerning animal control and natural resource conservation.
2.6.		

2.6.1.	The food safety plan shall address soil amendment risk, preparation, use, and storage.	If animal-based soil amendments or biosolids are used, records of composition, dates of treatment, methods utilized and application dates must be documented. Evidence of processing adequate to eliminate pathogens of human concern, such as letter of guarantee, certificate of analysis (COA) or any test results or verification data (e.g., time and temperature) demonstrating compliance with process or microbial standards, shall be documented. Such soil amendments must be produced and applied in accordance with applicable federal, state, or local regulations.
2.6.2.	If a soil amendment containing raw or incompletely treated manure is used, it shall be used in a manner so as not to serve as a source of contamination of produce.	If such a product is used, there shall be documentation of the composition, and time and method of application. Such use will be consistent with current industry practices or regulatory restrictions for that commodity.
2.7.		
2.7.1.	Equipment, vehicles, tools utensils and other items or materials used in farming operations that may contact produce are identified.	Operation maintains a list of equipment, vehicles, tools, utensils and other items or materials that may pose a risk of produce contamination during normal use.
2.7.2.	Equipment, vehicles, tools and utensils used in farming operations which come into contact with product are in good repair, and are not a source of contamination of produce.	The operation shall develop, implement, and schedule repair, cleaning, sanitizing, storage and handling procedures of all food contact surfaces to reduce and control the potential for contamination. As necessary for food safety, vehicles and equipment shall be properly calibrated, operated, maintained, and used as intended. Equipment traffic flow is prevented from traveling through an untreated manure area into the harvesting field. These procedures shall be documented. Product contact tools, utensils and equipment shall be made of materials that can be cleaned and sanitized. Procedures include equipment and vehicles that are in the field infrequently.
2.7.3.	Vehicles, equipment, tools and utensils shall be controlled so as not to be a source of chemical hazards.	Operation shall have a written procedure to address the spills and leaks (fuel, oil, hydraulic fluids) which might occur during equipment operation in the field.

2.7.4.	Vehicles, equipment, tools and utensils shall be controlled so as not to be a source of physical hazards.	Operation has a glass and brittle plastic policy that addresses glass on production equipment and in growing area. Inspections performed in compliance with the policy shall be documented.
2.7.5.	Cleaning and sanitizing procedures do not pose a risk of product contamination.	Equipment cleaning and sanitizing operations shall be conducted away from the product and other equipment to reduce the potential for contamination. Water used for cleaning and sanitizing shall meet the microbial standards for drinking water.
2.7.6.	Water tanks are cleaned at a sufficient frequency so as not to be a source of contamination.	There shall be a written procedure for cleaning water tanks, such as those used for dust control, the water from which may contact produce in the field.
3.		
3.1.		
3.1.1.	A preharvest risk assessment shall be performed.	The Operation shall have a preharvest assessment procedure, which describes when the assessment is performed and that it includes an evaluation of conditions that may be reasonably likely to result in physical, chemical, or biological contamination of the produce, and demonstrates that the operation is in compliance with the food safety plan. Results of the evaluation shall be documented.
3.2.		
3.2.1.	Operation has procedures for water used in contact with product or food contact surfaces.	Standard Operating Procedures (SOPs), including water-change schedules, shall be developed for all uses of water. Microbial and/or physical/ chemical (e.g., test strips) testing shall be performed, as appropriate to the specific operation, to demonstrate that acceptance criteria have been met.

3.2.2.	Water use SOPs address the microbial quality of water or ice that directly contacts the harvested crop or is used on food-contact surfaces.	If water or ice directly contacts the harvested crop or is used on food-contact surfaces, such as in the field, as the final wash step prior to consumer packaging, or as a cooling aid in a consumer package, operation's water use SOP requires that water or ice when applied meets the microbial standards for drinking water, as defined by prevailing regulation. Water may be treated (e.g., with chlorine) to achieve the microbial standards or to prevent cross-contamination. Ice and water shall be sourced/manufactured, transported, and stored under sanitary conditions. Special considerations or variances may be appropriate for some crops, e.g. cranberries and watercress, where deliberate flooding of the field is part of production and harvest practices.
3.2.3.	Water use SOPs address treatment of re-circulated water, if used.	Operation's water use SOPs require re-circulated water to be treated using an approved antimicrobial to prevent it from becoming a source of contamination, according to prevailing regulation or industry specific standards for the commodity.
3.2.4.	Water use SOPs address condition and maintenance of water-delivery system.	The water-delivery system shall be maintained so as not to serve as a source of contamination of produce, water supplies or equipment with pathogens, or to create an unsanitary condition.
3.2.5.	If applicable to the specific commodity, water use SOPs address control of wash water temperature.	For produce demonstrated as being susceptible to microbial infiltration from wash water, wash water temperature differentials during immersion shall be considered.
3.3.		
3.3.1.	Operation has written policy regarding storage of harvesting containers.	Harvesting containers shall be stored in a manner so as not to serve as a source of contamination to the extent feasible and appropriate.
3.3.2.	Operation has written policy regarding inspection of food contact containers prior to use.	Food-contact totes, bins, packing materials, other harvest containers, and pallets shall be visually inspected, clean, intact and free of any foreign materials prior to use. Containers shall be sufficiently maintained so as not to become a source of contamination.
3.3.3.	Operation has written policy regarding acceptable harvesting containers.	The types and construction of harvest containers and packing materials shall be appropriate to the commodity being harvested and suited for their intended purpose.
3.3.4.	Operation has written policy prohibiting use of harvest containers for non-harvest purposes.	Food-contact totes, bins and other harvest containers designated for harvesting shall not be used for other purposes unless clearly marked or labeled for that purpose.

3.4.			
	3.4.1.	Operation shall have a written policy that damaged or decayed produce is not harvested, or is culled.	Employees are trained that only sound produce appropriate for the intended use is harvested, and that produce that has been damaged to an extent that it poses a microbial food safety hazard is not harvested or is culled.
	3.4.2.	Product that contacts the ground shall not be harvested unless the product normally grows in contact with the ground.	Operation has considered and developed written policies regarding produce that comes in contact with the soil (e.g., drops). Policy shall be consistent with industry standards or prevailing regulations.
	3.4.3.	Harvest procedures shall include measures to inspect for and remove physical hazards.	Operation shall have procedures to detect glass/plastic breakage and remove possible physical contamination such as glass, metal, rocks, or other hazardous items, during harvesting operations.
	3.4.4.	Cloths, towels, or other cleaning materials that pose a risk of cross-contamination shall not be used to wipe produce.	Operations shall not use cloths or other cleaning materials to clean produce, unless there is a procedure to prevent cross-contamination.
	3.4.5.	Packaging materials shall be appropriate for their intended use.	The product contact packaging shall be appropriate to the commodity being harvested and suited for its intended purpose.
	3.4.6.	Packaging shall be stored in a manner that prevents contamination.	Packaging storage shall be designed to maintain packaging dry, clean and free from dirt or residues so it remains fit for the purpose. Particular care shall be taken to prevent packaging from becoming a harborage for rodents and other vermin. Packaging shall be stored separately from hazardous chemicals, toxic substances and other sources of contamination.
	3.4.7.	Operation has written policy regarding whether packing materials are permitted in direct contact with the soil.	If produce is packed in field, operation has considered and developed written policies regarding placement of packing materials directly on the soil, or whether a physical buffer (e.g., buffer bin or slip sheet) is required. Policy shall be consistent with industry standards.
3.5.			
	3.5.1.	Harvested produce is handled in a manner such that it is not likely to become contaminated.	Operation has a policy, in compliance with current industry practices or regulatory requirements for that commodity, regarding handling, walking, stepping, or lying on harvested produce, food contact surfaces or packaging materials, that may result in contamination.

3.5.2.	Materials that come in contact with the produce shall be clean and in good repair.	Operation has a policy that pallets, produce bins, totes and materials that come in contact with the produce or the containers during handling or storage shall be cleaned and, if practicable, sanitized sufficient so as not to be a source of contamination.
3.5.3.	Harvested produce shall be stored separately from chemicals which may pose a food safety hazard.	Chemicals, including cleaning and maintenance compounds shall be stored in an area separate from harvested produce.
4.		
4.1.		
4.1.1.	The operation shall have a policy, written procedures, and a checklist to verify cleanliness and functionality of shipping units (e.g., trailer).	Shipping units shall be clean, functional and free of objectionable odors before loading, in compliance with current industry practices or regulatory requirements for that commodity. Refrigeration units, if used, must be in working order.
4.1.2.	Loading/unloading procedures and equipment shall minimize damage to and prevent contamination of produce.	Personnel responsible for the loading and unloading of produce shall take steps to minimize the potential of physical damage to produce, which can introduce and/or promote the growth of pathogens. Loading/unloading equipment shall be clean and well maintained and of suitable type to avoid contamination of the produce.
4.1.3.	Trash shall not come in contact with produce.	The operation shall have a procedure describing how trash shall be handled and transported out of the field in a manner that does not pose a food safety risk.

Addendum 2: United Fresh Post Harvest Harmonized Standard

(Reproduced by permission—for brevity's sake only the requirement and procedure are reproduced here.)

	Requirement	Procedure
1.	**General Questions**	
1.1.	Management Responsibility	
1.1.1.	A food safety policy shall be in place.	A written policy shall outline a commitment to food safety, in general terms, how it is implemented and how it is communicated to employees, and be signed by Senior Management.
1.1.2.	Management has designated individual(s) with roles and responsibilities for food safety functions.	The Food Safety Plan shall designate who has the responsibility and authority for food safety, including a provision for the absence of key personnel. Twenty-four hour contact information shall be available for these individuals in case of food safety emergencies. These roles and responsibilities shall be communicated within the organization.
1.1.3.	There is a disciplinary policy for food safety violations	There shall be a policy that establishes corrective actions for personnel who violate established food safety policies or procedures.
1.2.	Food Safety Plan or Risk Assessment	
1.2.1.	There shall be a written Food Safety Plan. The plan shall cover the operation. The operation and products covered shall be defined.	The Food Safety Plan shall identify all locations of operation covered by the plan and shall identify physical, chemical, and biological hazards reasonably likely to occur and hazard control procedures, including monitoring, verification and recordkeeping, for all provisions covered by this audit.
1.2.2.	The Food Safety Plan shall be reviewed at least annually.	Operation shall be responsible for reviewing their Food Safety Plan at least annually, documenting the review procedure and revising the plan as necessary.
1.3.	Raw Material Sourcing	

1.3.1.	Operation has an Approved Supplier program for all incoming materials, including packaging.	Operation has and maintains a current list of approved raw material suppliers. Approved Supplier program includes a procedure for accepting materials from alternate sources.
1.3.2.	The Operation has a policy and takes affirmative steps to ensure that all fresh produce that are packed or stored in the facility are grown following requirements in *Field Operations and Harvesting* harmonized standard.	The Operation requires all raw product suppliers to provide evidence of food safety/GAP programs and compliance. Such evidence must include sufficient documentation to demonstrate that the supplier complies with the requirements in *Field Operations and Harvesting* harmonized standard.
1.4.	Documentation & Recordkeeping	
1.4.1.	Documentation shall be kept that demonstrates the Food Safety Plan is being followed.	Documents and records of procedures, standard operating procedures (SOPs) and policies shall be in place for meeting each of the food safety standards identified in the Food Safety Plan.
1.4.2.	Documentation shall be readily available for inspection.	Documents and records may be maintained on-site or at an off-site location, or accessible electronically (e.g., MSDS), and shall be available for inspection in a reasonable timeframe or as required by prevailing regulation.
1.4.3.	Documentation shall be retained for a minimum period of two years, or as required by prevailing regulation.	Document and record handling policy or procedures require that documentation required by the Food Safety Plan shall be retained for a minimum of two years, or as required by prevailing regulation.
1.5.	Worker Education and Training	

1.5.1.	All personnel shall receive food safety training.	All personnel shall receive training in the food safety policy and plan, food safety procedures, sanitation and personal hygiene appropriate to their job responsibilities. Personnel shall receive training at hire and refresher training at prescribed frequencies. Documentation of training is available.
1.5.2.	Personnel with food safety responsibilities shall receive training sufficient to their responsibilities.	The individual designated for food safety responsibilities demonstrates knowledge of food safety principles. Food safety designate has completed at least one formal food safety course/workshop or by job experience.
1.6.	Traceability	
1.6.1.	A documented traceability program shall be established.	Records that enable reconciliation of product delivered to recipients (one step forward) shall be maintained except for direct to consumer sales. Records shall be maintained that link product with source of the produce and other supplies and raw materials (one step backward). Records shall include the items and date of receipt, lot numbers, quantities, source of the produce, and transporter. Additional information may be included. Contents and retention of records shall be consistent with applicable regulations.

1.6.2.	A trace back and trace forward exercise shall be performed at least annually.	The trace back and trace forward exercise shall achieve accurate traceability within 4 hr or as required by applicable regulations. Trace exercise shall achieve 100% reconciliation of product to recipients.
1.7.	Recall Program	
1.7.1.	A documented recall program, including written procedures, shall be established.	The recall program shall have a designated recall team. A mock recall exercise shall be performed at least annually at the Operation being audited. The mock recall shall include the trace back and trace forward exercise and shall be completed as stated in the program and in compliance to applicable regulations.
1.8.	Corrective Actions	
1.8.1.	The Operation shall have documented corrective action procedures.	A documented Corrective Action is required for an observation or audit that contains a non-conformance with food safety requirements. The responsibility, methods, and timelines to address Corrective Actions shall be documented and implemented.
1.9.	Self-audits	
1.9.1.	The Operation shall have documented self-audit procedures.	Internal audits will be conducted at a minimum annually by an assigned individual who is knowledgeable in this standard, utilizing this standard to assist in the self-audit. All aspects of the Operation's Food Safety Plan will be audited and a written record of required corrective action will be documented.
1.10.	Agricultural Chemicals /Plant Protection Products	

1.10.1.	Use of agricultural chemicals shall comply with label directions and prevailing regulation.	Agricultural chemicals applied post-harvest (e.g., biocides, waxes and plant protection products) must be registered for such use as required by prevailing regulation, and used in accordance with label directions including application rates, worker protection standards, personal protection equipment, container disposal, storage, and all requirements specified for the chemical or compound. Records of use are maintained.
1.10.2.	If product is intended for export, pre- and post-harvest agricultural chemical use shall consider requirements in the intended country of destination.	The operation shall have procedures, such as review of pre-harvest intervals and adjustment of post-harvest application rates, sufficient to meet the MRL entry requirements of the country(ies) in which the product is intended to be traded, if known during post-harvest handling.
1.10.3.	Agricultural chemicals shall be applied by trained, licensed or certified application personnel, as required by prevailing regulation.	Operation maintains records demonstrating that all personnel responsible for chemical applications are trained and/or licensed, or supervised by licensed personnel, in compliance with prevailing regulation.
1.11.	Water/Ice	

1.11.1.	Water use SOPs address the microbial quality of water or ice that directly contacts the harvested crop or is used on food-contact surfaces.	If water or ice directly contacts the harvested crop or is used on food-contact surfaces, Operation's water use SOP requires that water or ice when applied meets the microbial standards for drinking water, as defined by prevailing regulation or the country in which the product is intended to be traded, whichever is more stringent. Water may be treated (e.g., with chlorine) to achieve the microbial standards or to prevent cross-contamination. Ice and water shall be sourced/manufactured, transported, and stored under sanitary conditions.
1.11.2.	A water system description shall be prepared.	Water sources and the operations they serve shall be documented and current. The description shall include one or more of the following: maps, photographs, drawings (hand drawings are acceptable) or other means to communicate the location of water source(s), permanent fixtures and the flow of the water system (including holding systems, reservoirs or any water captured for re-use). Permanent fixtures include wells, gates, reservoirs, valves, returns, backflow prevention and other above ground features that make up a complete water distribution system shall be documented in such a manner as to enable location in the operation.

1.11.3.	Documented scheduled assessment of water system including delivery equipment shall be performed.	The water-delivery system shall be maintained so as not to serve as a source of contamination of produce, water supplies or equipment with pathogens, or to create an unsanitary condition. Water installations and equipment are constructed and maintained to prevent back siphonage backflow and cross connections between product contact water and waste water. Routine checks verify that back siphonage and backflow prevention units are functioning properly (annual or as needed to maintain continuous protection). Results are documented.
1.11.4.	The sewage disposal system is adequate for the process and maintained to prevent direct or indirect product contamination.	The human waste and gray water sewage system has sufficient capacity to handle the operation's peak flows and not cause direct or indirect product contamination. Cross-connections with product contact water systems are prohibited.
1.11.5.	Water-change schedules shall be developed for all uses of water where water is re-used.	Operation shall have procedures for changing water that is re-used, such as recirculated water, flumes and dump tanks.
1.11.6.	Re-circulated water that contacts product or food contact surfaces shall be treated using an approved antimicrobial process or chemical treatment.	Re-circulated water shall be treated using an antimicrobial treatment sufficient to prevent cross-contamination. Treatments shall be in compliance with prevailing regulation or the country in which the product is intended to be traded, whichever is more stringent.

1.11.7.	If used, water antimicrobial treatments shall be monitored sufficiently to assure continuous control.	Microbial, physical or chemical testing shall be performed, as appropriate to the specific operation, to demonstrate that acceptance criteria have been met.
1.11.8.	If applicable to the specific commodity, water use SOPs address control of immersion water temperature.	For produce that is immersed in water and demonstrated as being susceptible to microbial infiltration from water, water temperature differentials during immersion shall be controlled in accordance with prevailing regulation or industry guidelines.
1.12.	Containers, Bins	
1.12.1.	Operation has written policy regarding storage and post-storage handling of product-contact containers.	Product-contact containers, as appropriate to the specific operation (e.g., harvest bins, totes, crates, sacks, buckets, finished product clam shells, bags or packaging films), shall be stored, or handled (e.g., cleaned prior to post-storage use), in a manner so as not to serve as a source of contamination
1.12.2.	Operation has written policy regarding whether product-contact containers are permitted in direct contact with the ground.	If produce does not normally contact the ground during production, Operation has considered and developed written policies regarding placement of product-contact containers directly on the ground, or whether a physical buffer (e.g., buffer bin or slip sheet) is required, or use of containers constructed to prevent contact of the produce or produce contact surfaces with the ground. Policy shall be consistent with industry standards.

1.12.3.	Operation has written policy regarding inspection of food contact containers and bins prior to use.	Food-contact totes, bins, packing materials, other harvest containers, and pallets shall be visually inspected, clean, intact and free of any foreign materials prior to use. Containers shall be sufficiently maintained so as not to become a source of contamination.
1.12.4.	Operation has written policy regarding acceptable product-contact containers.	The types and construction of product-contact containers and packing materials shall be appropriate to the commodity being handled and suited for their intended purpose. Produce shall only be stored in clean and sanitary containers.
1.12.5.	Operation has written policy prohibiting use of product-contact containers for non-product purposes unless clearly marked or labeled for that purpose.	Food-contact totes, bins and other product-contact containers shall not be used for other purposes unless the Operation has a policy or procedure that clearly designates approved non-product contact uses and how the containers are to be marked or labeled for that purpose. Food-contact totes, bins and other packing containers and equipment that are no longer cleanable shall not be used for packing but can be used for other non-food uses if clearly marked/labeled.
1.12.6.	Pallets shall be kept clean and in good condition as appropriate for their intended use.	Operation inspects pallets prior to use for conditions that may be a source of produce contamination. Pallets that are not cleanable are removed from use. Pallets and other wooden surfaces are properly dried after being washed.
1.13.	Facility, Equipment, Tools	

1.13.1.	Facility shall be designed, constructed and maintained in a manner that prevents contamination of produce during staging and cooling.	Facility and equipment structures and surfaces (floors, walls, ceilings, doors, frames, hatches, etc.) shall be constructed in a manner that facilitates cleaning and sanitation and does not serve as harborage for contaminants or pests. Chill and cold storage loading dock areas shall be appropriately sealed, drained and graded. Fixtures, ducts, pipes and overhead structures shall be installed and maintained so that drips and condensation do not contaminate produce, raw materials or food contact surfaces. Water from refrigeration drip pans shall be drained and disposed of away from product and product contact surfaces. Drip pans and drains shall be designed to assure condensate does not become a source of contamination. Air intakes shall not be located near potential sources of contamination.
1.13.2.	A Preventive Maintenance and/or Master Cleaning Schedule, with related SOPs, shall be established	There is a written cleaning and sanitation schedule for all food and non-food contact surfaces including floors, drains, walls, ceilings and other surfaces that may pose a source of product contamination. Roof leaks shall be promptly identified, controlled and repaired. Operation has procedures for cleaning and sanitation of cooling equipment. Drip pans and drains shall be maintained to assure condensate does not become a source of contamination.

1.13.3.	All cleaning agents shall be approved for their intended use on food contact surfaces.	All chemicals used for cleaning or sanitizing of food contact equipment, tools, utensils, containers and other food contact surfaces shall be approved for that use, according to the chemical manufacturer or supplier and all federal, state and local requirements, and shall be used in a manner consistent with the approved use.
1.13.4.	Cleaning equipment and tools are clean, in working order and stored properly away from product handling areas.	Equipment, utensils and tools used for cleaning or sanitizing, including food contact and non-food contact surfaces, are maintained in a manner sufficient to avoid becoming a source of produce contamination and are stored away from product handling areas.
1.13.5.	Food contact surfaces shall be cleaned, sanitized and maintained according to the Food Safety Plan	Prior to use, the lines used for washing, grading, sorting, or packing shall be cleaned and sanitized as appropriate per risk assessment. When in use, the lines shall be maintained so as not to be a source of contamination with pathogens.
1.13.6.	Adequate lighting shall be provided in all areas.	Lighting in all areas shall be sufficient to enable cleaning, sanitation, repairs, etc.
1.13.7.	Where temperature control is required for food safety, cooling facilities shall be fitted with temperature monitoring equipment or suitable temperature monitoring device.	Temperature monitoring equipment shall be located in all temperature controlled areas, and shall be located so as to accurately monitor the temperature. Temperature measuring devices shall be monitored and calibrated on a scheduled basis or as needed.

1.13.8.	Cooling equipment shall be maintained so as not to be a source of product contamination.	Cooling equipment (e.g. hydrocoolers, air coolers), shall be inspected, all debris removed, and cleaned and sanitized according to written sanitation SOPs.
1.13.9.	Transporting equipment shall be maintained to prevent contamination of products being transported.	Pallet jacks, carts, trolleys and forklifts, shall be maintained to prevent contamination of products being transported and are listed on the Preventive Maintenance and/or Master Cleaning Schedules.
1.13.10.	Outside garbage receptacles/dumpsters are closed and located away from facility entrances and the area around such sites is reasonably clean.	Waste containers and compactors are located away from produce handling areas, are closed or have lids (except for waste collection/cull trailers in active use), are emptied on a scheduled basis or as needed, and weeds and other pest harborage are minimized around the containers.
1.13.11.	The plant grounds are reasonably free of litter, waste culls, vegetation, debris and standing water.	Operation has procedures to maintain the grounds surrounding the building in a manner to minimize sources of contamination, such as litter, vegetation, waste culls, debris and standing water that may be pest attractants or harborages. Vegetation that does not serve as an attractant or harborage is permitted.
1.14.	Storage	
1.14.1.	Product storage areas and conditions shall be appropriate to the commodities stored.	Produce storage locations and conditions shall not pose a risk of produce contamination, consistent with industry standards or prevailing regulation.
1.14.2.	Iced produce is handled so as not to serve as a source of contamination.	Protective measures are provided in areas where iced product is stored over food items in order to prevent melting ice from contaminating product below.

1.14.3.	Non-product storage areas shall be maintained so as not to be a source of product or materials contamination.	Areas designated to store materials, whether indoors or out, shall be clean, well ventilated, and designed to protect materials and produce from contaminants.
1.14.4.	Materials and packaging materials shall be protected from contaminants.	Materials stored in uncovered areas shall be protected from condensate, sewage, dust, dirt, chemicals, allergens or other contamination. Materials shall be stored off the floor/ground on pallets, slip-sheets or stands and covered where applicable.
1.14.5.	Adequate space shall be maintained between rows of stored materials to allow cleaning and inspection.	Materials shall be stored away from walls and ceilings. Written procedures shall be followed to guarantee the proper cleaning, inspection and monitoring for pest activity in storage areas.
1.14.6.	All chemicals shall be stored in a secure separate area. All chemicals shall be properly labeled.	Chemicals, including cleaning and maintenance compounds and lubricants, when not being used, are stored away from product handling areas and in a manner that inhibits unauthorized access. Food-grade and non food-grade lubricants are kept separate from each other.
1.15.	Waste Material	
1.15.1.	Waste materials and their removal are managed to avoid contamination.	Trash, leaves, trim, culls, waste water and other waste materials are removed from the produce handling areas at a frequency sufficient to avoid becoming a source of produce contamination.
1.16.	Outside Grounds	

1.16.1.	Operation has procedures to prevent pest harborage in any equipment stored near the building.	Equipment stored outside is stored away from the building perimeter. Equipment is not to accumulate near the building. Bone yards are located away from the building. Outside equipment storage areas are included in pest control program.
1.17.	Glass Control	
1.17.1.	Only essential glass and brittle plastic shall be present in the facility.	Light bulbs, fixtures, windows, mirrors, skylights and other glass and brittle plastic in the facility or in the product path entering or exiting the facility shall be of the safety type, or shall be otherwise protected to prevent breakage. If glass or brittle plastic must be used, there shall be a written glass and brittle plastic control policy, including a glass and brittle plastic register.
1.18.	Leaks/Lubrication	
1.18.1.	Equipment lubrication is managed so as not to contaminate food products.	Only food-grade lubricants are used on food processing and packaging equipment, or on any other equipment where incidental food contact may occur, unless the equipment manufacturer specifies only a non-food grade lubricant. Lubricant leaks are fixed or catch pans are installed to prevent product contamination.
1.19.	Equipment and Utensil Construction	
1.19.1.	All food contact equipment, tools and utensils are designed and made of materials that are easily cleaned and maintained.	The Operation shall develop, implement, and schedule repair, cleaning, sanitizing, storage and handling procedures of all food contact surfaces to reduce and control the potential for contamination. These procedures shall be documented. Product contact tools, utensils and equipment shall be made of materials that can be cleaned and sanitized.

1.19.2.	Equipment is installed in a way that provides access for cleaning.	Cooling, packing and other food contact equipment is installed away from walls and otherwise positioned so as not to inhibit access for proper cleaning.
1.19.3.	Catwalks above product zones are protected to prevent produce or packaging contamination.	Where workers walk over product contact surfaces, those walkways are solid surface or have catch trays installed, are protected by kick plates, product covers or other barriers.
1.20.	Temporary Repairs	
1.20.1.	Any temporary repairs on food contact surfaces are constructed of food-grade material. Operation has a procedure to ensure that permanent repairs are implemented in a timely manner.	Operation has procedures to ensure temporary repairs are compliant with all food safety requirements, and do not create potential sources of chemical, microbiological or physical contamination. Permanent repairs are implemented as soon as practical; Operation establishes timelines and responsibilities for completion.
1.21.	Worker Health/Hygiene and Toilet/Handwashing Facilities	
1.21.1.	Restrooms shall be designed, constructed, and located in a manner that minimizes the potential risk for product contamination.	Restrooms shall be designed and constructed in a manner that minimizes the potential risk for product contamination, are located away from produce handling areas, and are directly accessible for servicing.
1.21.2.	Toilet facilities shall be of adequate number, easily accessible to employees and in compliance with applicable regulation.	The Operation will have verification that the number of toilet facilities and their location relative to employees meets the more stringent of federal, state or local regulations.

1.21.3.	The practice of disposing of used toilet tissue on the floor, in trash receptacles, or in boxes is prohibited.	Operation shall instruct employees that used toilet tissue shall only be disposed of in the toilet.
1.21.4.	Toilet and hand wash stations shall be maintained in a clean and sanitary condition.	Toilet paper shall be available in toilet facility. Restrooms shall include hand wash facilities with water that meets the microbial standard for drinking water, hand soap, disposable towels or other hand drying device, and towel disposal container. Gray water is plumbed or captured for disposal.
1.21.5.	Signage requiring handwashing is posted.	Signage in applicable languages and/or pictures shall be provided adjacent to hand wash facilities requiring people to wash their hands after each toilet visit.
1.21.6.	When appropriate, racks and/or storage containers or designated storage area for protective clothing and tools used by employees shall be provided.	When employees wear protective clothing, such as aprons and gloves, the Operation shall have a policy that the clothing not be left on product, work surfaces, equipment or packaging material but hung on apron and glove racks provided. Racks shall be available and located so as to avoid potential contamination. In addition, storage containers or designated storage areas shall be provided to ensure tools used by employees are properly stored prior to entering toilet facilities.
1.21.7.	Employees and visitors shall follow all personal hygiene practices as designated by the Operation.	Operation's hygiene policies shall apply to all employees, contractors, visitors, buyers, product inspectors, auditors, and other personnel in the facility. The Operation shall designate competent supervisory personnel to ensure compliance with the requirements in this section.

1.21.8.	Workers and visitors who show signs of illness shall be restricted from direct contact with produce or food-contact surfaces.	Operation shall have a policy that restricts employees, contractors, visitors, buyers, product inspectors, auditors, and other personnel in the facility who show signs of illness (e.g., vomiting, jaundice, diarrhea) from contact with product or food contact surfaces.
1.21.9.	Personnel with exposed cuts, sores or lesions shall not be engaged in handling product.	Minor cuts or abrasions on exposed parts of the body are acceptable if covered with a non-permeable covering, bandage or glove. Bandages on hands shall be covered with gloves in compliance with Operation's glove policy.
1.21.10.	First aid kits shall be accessible to all personnel.	The kits shall be readily available in the facility and maintained in accordance with prevailing regulation. The kit materials shall be within shelf life and kept in a sanitary and usable condition.
1.21.11.	Smoking, chewing, eating, drinking (other than water), chewing gum and using tobacco shall be prohibited except in clearly designated areas.	Operation shall have policy prohibiting smoking, eating, chewing gum or tobacco, drinking other than water except in designated areas. Such areas shall be designated so as not to provide a source of contamination.

1.21.12.	Personnel shall be required to wash their hands before beginning or returning to work, after each visit to the toilet and whenever their hands may have become a source of contamination.	Personnel shall wash their hands prior to start of work, after each visit to a toilet, after using a handkerchief/tissue, after handling contaminated material, after smoking, eating or drinking, after breaks and prior to returning to work and at any other time when their hands may have become a source of contamination.
1.21.13.	If gloves are used, the Operation shall have a glove use policy.	If rubber, disposable, cloth or other gloves are used in contact with product, the Operation shall have a glove use policy that specifies types of glove materials that are allowed, how and when gloves are to be used, cleaned, replaced and stored. Policy shall be in compliance with current industry practices or regulatory requirements for that commodity.
1.21.14.	Clothing, including footwear, shall be effectively maintained, stored, laundered and worn so as to protect product from risk of contamination.	Operation shall have a policy that employee clothing shall be clean and appropriate for the operation.
1.21.15.	If protective clothing is required by the Operation in product handling areas, it shall be handled in a manner to protect against contamination.	Protective clothing, such as aprons and gloves, shall not be left on product, work surfaces, equipment or packaging material but hung on apron and glove racks or in designated areas. Operation shall have a policy regarding whether protective clothing can be taken home.

1.21.16.	The use of hair coverings shall be in compliance to company policy and applicable regulation.	The Operation shall have a policy that addresses use of hair coverings (e.g., hair nets, beard nets, caps), which is in compliance with prevailing regulation.
1.21.17.	The wearing of jewelry, body piercings and other loose objects (e.g. false nails) shall be in compliance to company policy and applicable regulation.	Operation shall have a policy to minimize risk for jewelry or loose objects to be a source of product contamination. Policy shall be in compliance with current industry practices or regulatory requirements for that commodity.
1.21.18.	Employees' personal belongings shall be stored in designated areas.	Operation shall have a policy for when and how employee's personal belongings shall be stored so as not to be a source of product contamination.
1.21.19.	Break areas shall be designated and located away from food contact/handling zones.	Operation shall have a written policy that break areas are located so as not to be a source of product contamination.
1.22.	Temperature Control	
1.22.1.	When produce is cooled, it is cooled to temperatures appropriate to the commodity according to current established regulatory or industry standards.	When required for food safety or by industry guidelines, steps are taken to minimize temperature increases and minimize the time between produce receipt and cooling at the operation. The product temperature and equipment control mechanisms are calibrated and monitored at a defined frequency and temperatures are kept appropriate to the commodity. Records are maintained.
1.23.	Packing and Handling	

1.23.1.	If applicable, Operation has a written Allergen Control Program	The Allergen Control Program lists the allergens in use or storage at the facility specific to country regulations. If applicable, procedures address identification and segregation of allergens during storage and handling as based on a risk assessment conducted by the facility
1.23.2.	Specifications for all packaging materials that impact on finished product safety and quality shall be provided and comply with prevailing regulations.	The methods and responsibility for developing and approving detailed specifications and labels for all packaging shall be documented. A register of packaging specifications and label approvals shall be maintained and kept current.
1.24.	Pest and Animal Control	
1.24.1.	Operation has procedures to manage pests to the extent appropriate to the facility.	Operation has a written pest control program, performed by a trained pest control operator (or licensed where required by prevailing regulation). The written program includes policies and procedures applicable to that operation, such as storage of outside equipment or other factors dealing with pest harborages, and maps of the location of pest traps outside and inside the facility. Operation maintains a pest-control log that includes dates of inspection, inspection reports and steps taken to eliminate any problems. Applications of pesticides (e.g., insecticides, rodenticides) shall be performed in compliance with local, state, and federal pesticide regulations.

1.24.2.	Operation restricts animals from food handling facilities.	Domestic animals are prohibited from pack house, cooling, and storage facilities unless procedures are in place for their safe presence. Procedures are in place to exclude wild and feral animals to the degree practical.
1.24.3.	If used, pest control devices including rodent traps and electrical flying insect devices, are located so as to not contaminate produce or food handling surfaces.	Only non-toxic traps and pest control devices are used inside the packing house or storage facility.
1.25.	Sampling/Testing	
1.25.1.	Where laboratory analysis is required in the Food Safety Plan, testing shall be performed by a GLP laboratory using validated methods.	Operation utilizes laboratories that have, at minimum, passed a Good Laboratory Practices (GLP) audit or participates in a Proficiency Testing program, and utilizes BAM, AOAC International or testing methods that have been validated for detecting or quantifying the target organism(s) or chemical(s).
1.25.2.	Where microbiological analysis is required in the Food Safety Plan, samples shall be in accordance with an established sampling procedure.	Operation utilizes a written sampling protocol when collecting samples for microbiological testing.
1.25.3.	Tests, their results and actions taken must be documented.	All results for microbiological testing required in the Operation's Food Safety Plan shall be recorded and the records maintained for two years.

1.25.4.	All required testing shall include test procedures and actions to be taken based on the results.	For all microbiological testing required by the Food Safety Plan, Operation has a written testing procedure that includes test frequency, sampling, test procedures, responsibilities and actions to be taken based on results. If finished product is tested for pathogens or other adulterants, Operation's procedures require that it shall not be distributed outside the Operation's control until test results are obtained.
2.	**Packinghouse**	
2.1.	Operation Food Safety Plan includes produce washing process, if used.	If produce is washed, an initial risk assessment of the washing process shall be performed that takes into consideration the commodity, type of wash system, type of sanitizer, and water quality.
2.2.	Debris and damaged produce shall be removed from wash areas/dump tanks to the extent possible.	Operation has procedures to determine how and when debris shall be removed from wash areas/dump tanks.
2.3.	Operation has documentation demonstrating regulatory approval of the wash water antimicrobials in use.	Only wash water antimicrobials or antimicrobial systems registered or approved by EPA, FDA or the prevailing regulatory agency for their specific intended use may be used in the dump tank wash water, on the spray line or other food contact purposes.
2.4.	If wash water antimicrobial is used, it shall be used in accordance with established operational procedure and manufacturer instructions.	Records shall be kept. Operation shall have a procedure that includes minimum limits for antimicrobial in wash water for food safety. Procedure shall include how to control, monitor and record use of wash water antimicrobial as needed to assure compliance with minimum limits. Operation shall have a procedure as to what corrective actions are taken if criteria are not met.

2.5.	All instruments used to measure temperature, pH, antimicrobial levels and or other important devices used to monitor requirements in this section shall be calibrated at a frequency sufficient to assure continuous accuracy.	Records shall be kept. If an ORP system is used, an independent measurement shall be used to verify compliance. Test methods or test strips used to monitor requirements shall be appropriate to their use and sufficiently sensitive to their intended purpose.
2.6.	Foreign material control devices are inspected and maintained	If included in the Food Safety Plan, foreign material control devices shall be included as part of a Preventive Maintenance Schedule or other program and maintained to ensure effective operation. Calibration checks shall be performed according to written procedure or manufacturer's recommendations.
3.	**Transportation (Packinghouse to Customer)**	
3.1.	Temperature Control (When refrigerated transport is required for food safety)	
3.1.1.	There is a written policy for transporters and conveyances to maintain a specified temperature(s) during transit.	Transporters have written, predetermined temperature ranges for commodities being transported.

3.1.2.	Prior to loading, the vehicle shall be pre-cooled.	The proper temperature for pre-cooling is appropriate to the type of produce and as specified by documented protocol.
3.1.3.	The refrigerated transport vehicles shall have properly maintained and fully functional refrigeration equipment.	Operation has a written policy that refrigerated transportation equipment shall be controlled by a thermostatic device as necessary to maintain temperatures in the cargo area for the particular type of produce being transported and as specified by documented protocol.
3.1.4.	Where required, temperatures of product are taken and recorded prior to or upon loading.	Operation has a written procedure for when and how to measure product temperatures prior to or during loading
3.2.	Equipment Sanitation and Maintenance	
3.2.1.	The Operation shall have a policy, written procedures, and a checklist to verify cleanliness and functionality of shipping units (e.g., trailer).	Shipping units shall be clean, functional and free of objectionable odors before loading, in compliance with current industry practices or regulatory requirements for that commodity. Refrigeration units, if used, must be in working order. Procedures include prohibition of raw animal or animal product transport, or other materials that reasonably may be a source of contamination with biological, chemical or physical hazards. A responsible individual shall sign or initial the completed checklist or inspection report.
3.2.2.	Loading/unloading procedures and equipment shall minimize damage to and prevent contamination of produce.	Personnel responsible for the loading and unloading of produce shall take steps to minimize the potential of physical damage to produce, which can introduce and/or promote the growth of pathogens. Loading/unloading equipment shall be clean and well maintained and of suitable type to avoid contamination of the produce.

| 3.2.3. | Trash shall not come in contact with produce. | Trash handling and removal shall not pose a hazard of contamination of produce. |

Form copyright 2012 by Equicert

Request for Audit—Copy and mail in

Name of Farm: _____

Telephone Number: _____

Audit Site Physical Address: _____
_____

Farm Mailing Address: _____

_____

_____

I request the following audit: (circle as many as apply):

Individual Farm GAP Audit   Certified Horsepowered Farm Audit   CanadaGAP   Produce Auction GHP Audit   Fossil Fuel Free Audit   Group GAP Audit   Group GAP Horsepowered Audit

I understand that I am requesting an audit that I may pass or fail. I understand that I am responsible to be ready for audit, and to pay for the audit whether I pass or fail. I understand that I am responsible for making sure that the audit requested meets buyer needs.

_____   _____
Name   Date

Send to:

Equicert
5467 S 800 W
Delphi, IN 46923-8823

www.ingramcontent.com/pod-product-compliance
Lightning Source LLC
Chambersburg PA
CBHW071808170526
45167CB00003B/1226